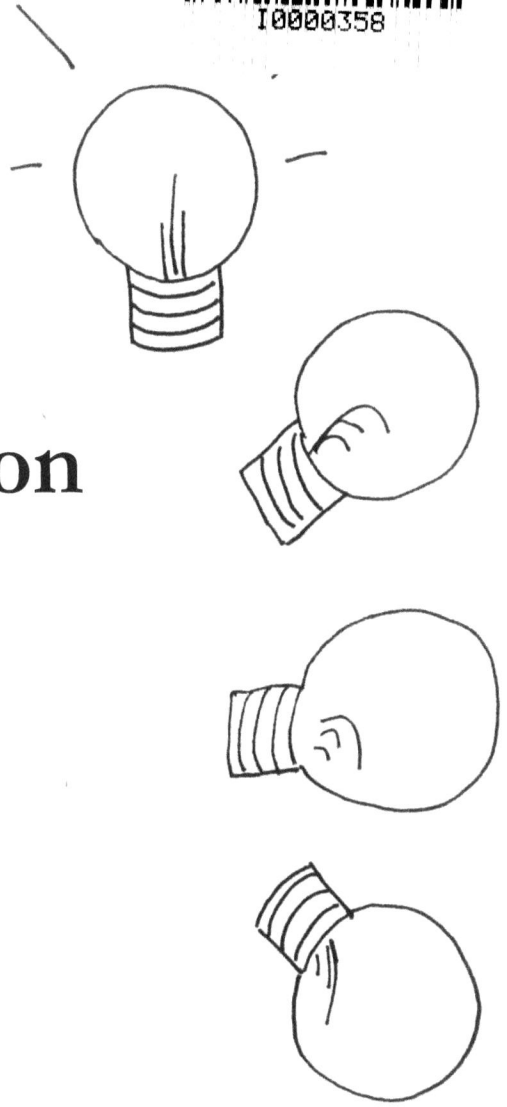

Impossible Evolution

A few problems with the Theory of Evolution

Renée Ellison

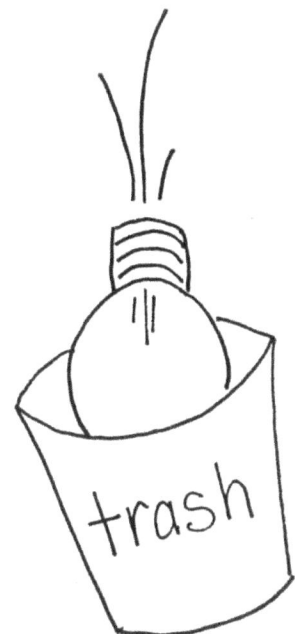

**"I am fearfully and wonderfully made.
Your works are wonderful;
I know that full well."
(Psalm 139:14)**

"The hour is short" (1 Corinthians 7:29)
"Work while it is day; night cometh when no man can work." (John 9:4)

———————————————

Library of Congress Cataloging-in-Publication Data

Ellison, Renée R.

Impossible evolution: A few problems with the theory of evolution.

Durango, Colo.: Homeschool How-Tos, 2018.

70 p.
Teach faster series.
Home schooling - Curricula.
Intelligent design (Teleology)
Evolution.
Evolution (Biology)
Religion and science.
Biology - Evolution vs. Intelligent Design.
Science - Study and teaching - Philosophy.

Printed in the United States of America.

———————————————————————————

ISBN-13: 978-0-9987894-2-2

Website: http://www.homeschoolhowtos.com

Email: info@homeschoolhowtos.com

This little book is in three parts:

Part 1 uses pictures to frame the debate.

Part 2 is a synopsis of problems and proofs.

(See page 13 for the table of contents of Part 2.)

Part 3 is a review of the debate via questions galore.

(See page 55 for the table of contents of Part 3.)

So what?

Does it really matter what we believe...

...about how the earth began?
...and how WE began?

Yes!

If we are the result of a random accident in the universe, *why live?*

If, on the other hand, we are part of God's creation,
we have great purpose!

"Big Bang"

"Big God"

VS.

"Big-Bang"
Believers
believe

↓

Life starts with a hunk of matter

"stuff"

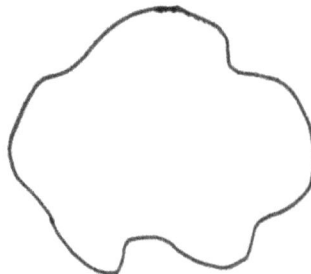

...and then takes millions of years to evolve into everything else.

"Big-God" Believers believe

↓

Life starts with information (codes like DNA)

imparted via fiat (language)

א ה

God spoke it into existence fully formed.

Starter problem #1 with evolution

Where and how did a hunk of "mere matter" acquire the intelligence to use a tool to make a tool?

No animal is able to do this even yet!

"...who teaches us more than He teaches the beasts of the earth, and makes us wiser than the birds in the sky?" Job 35:11

Starter problem #2 with evolution

Has any human being

ever been able

to make somethin'

out of nothing?

Or life from non-life?

Secret formula ....where did that original hunk come from?

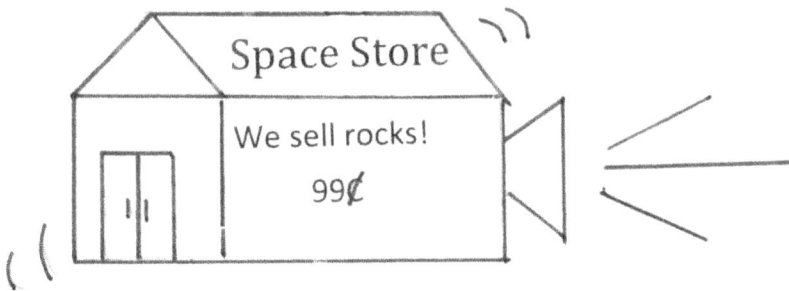

Space Store

We sell rocks!

99¢

Secret formula ...how did life spring from non-life?

dead-beat rock

speak rock!

NOT!

Sigh

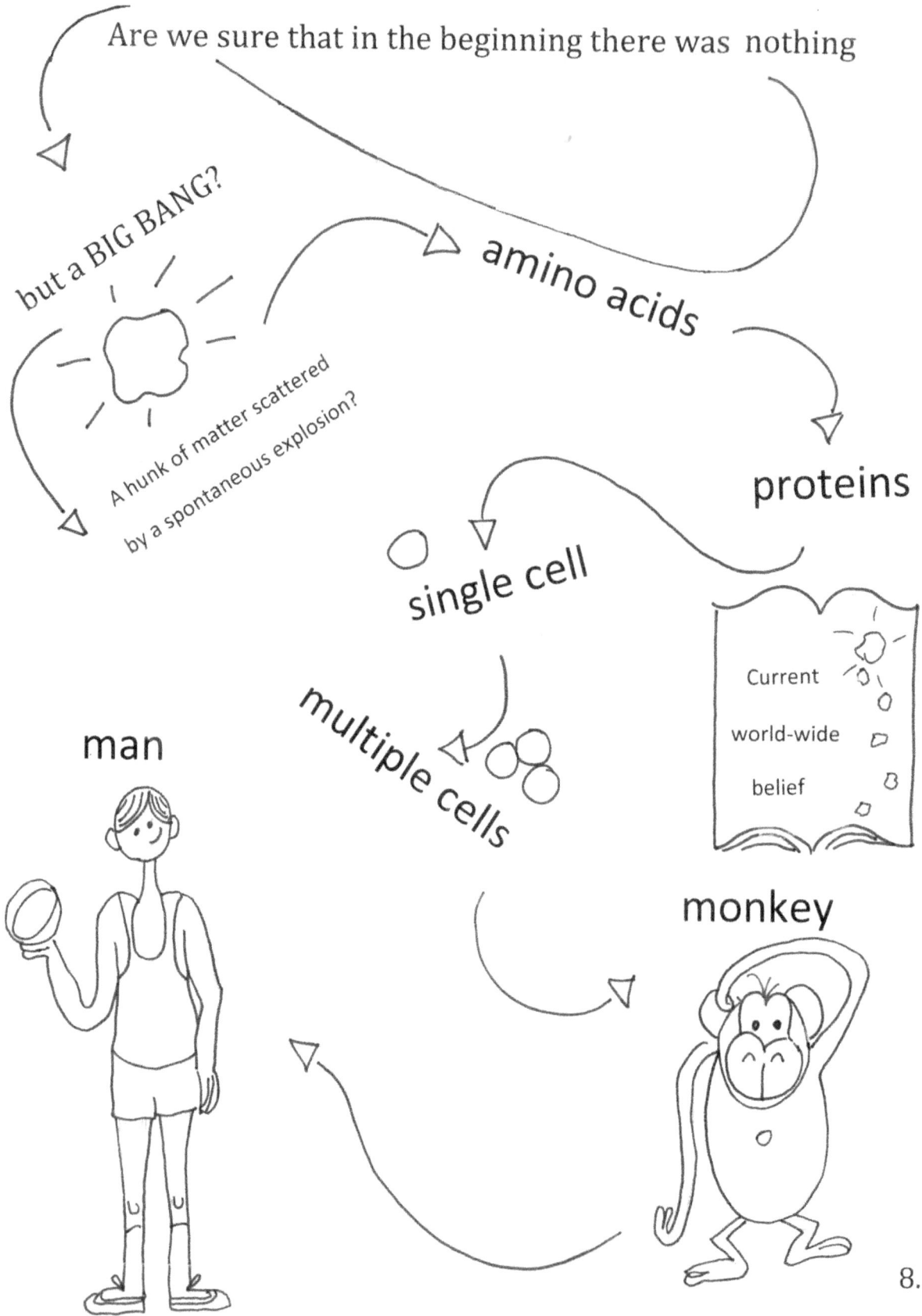

Are we sure that in the beginning there was nothing

but a BIG BANG?

amino acids

A hunk of matter scattered by a spontaneous explosion?

proteins

single cell

Current world-wide belief

multiple cells

man

monkey

But where did these come from?

Water? From dry matter? "Nothin' the "matter" with that!"

ABC's? Language? Talk? From rocks came books? "Sure!"

DNA? From chaos came formulas, physics, math? "OK!"

Intelligence? IQ's ? Genius'? Musicians? Chess? "Um-hmm!"

Conscience/Guilt/Justice? From the physical came also all the metaphysical? "Anything you say!"

The Periodic Table?

Did 103 such vastly different elements come from the same original source/matter? Copper? Tin? Nickel? Sulfur? Oxygen?

From 2 atoms came 103 different building materials ? Seriously? Really?

God...for whom and through whom everything exists. Hebrews 2:10
His works have been finished since the creation of the world. Hebrews 4:3b

He is the maker of all things.
Jeremiah 10:16

Has not my hand made all these things,
And so they came into being." Isa. 66:2

Levi was still in the body of his ancestor. Hebrews 7:10

For in Him all things were created;
Things in heaven and on earth, visible and invisible...
All things have been created through him and for him.
He is before all things, and in him all things hold together.
Colossians 1: 16, 17

The universe was formed by the word of God, so that what is seen was not made out of what was visible. Hebrews 11:3

God made the earth by His power
Founded the earth by his wisdom
Stretched out the heavens by his understanding;
When He thunders the waters in the heavens roar.
Jeremiah 10:12

I worship the Lord, the God of heaven, who made the sea and the dry land. Jonah 1:9

I made the sand a boundary for the sea,
An everlasting barrier it cannot cross.
The waves may roll, but they cannot prevail;
They may roar, but they cannot cross it.
Jeremiah 5:22

...his son...through whom also he made the universe...sustaining all things by his powerful word...Hebrews 1:2-3

..you laid the foundations of the earth, and the heavens are the work of your hands.
Hebrews 1:10

Satan is the father of lies....John 8:44

Is Scripture Wrong?

"By the **word** of the Lord were the heavens made;
And all the host of them by the **breath of his mouth**."
Psalm 33:6

Via the vapor of his breath he brought forth unfathomable multitudes of created things directly from his mind. Twas a stunning performance ...absolutely stunning.

A Dying Theory

Evolutionists blew "hot-air" upon God's magnificent and complex act of creation, happy to be rid of Him. But as we shall soon see, evolution is not only improbable, it is impossible!

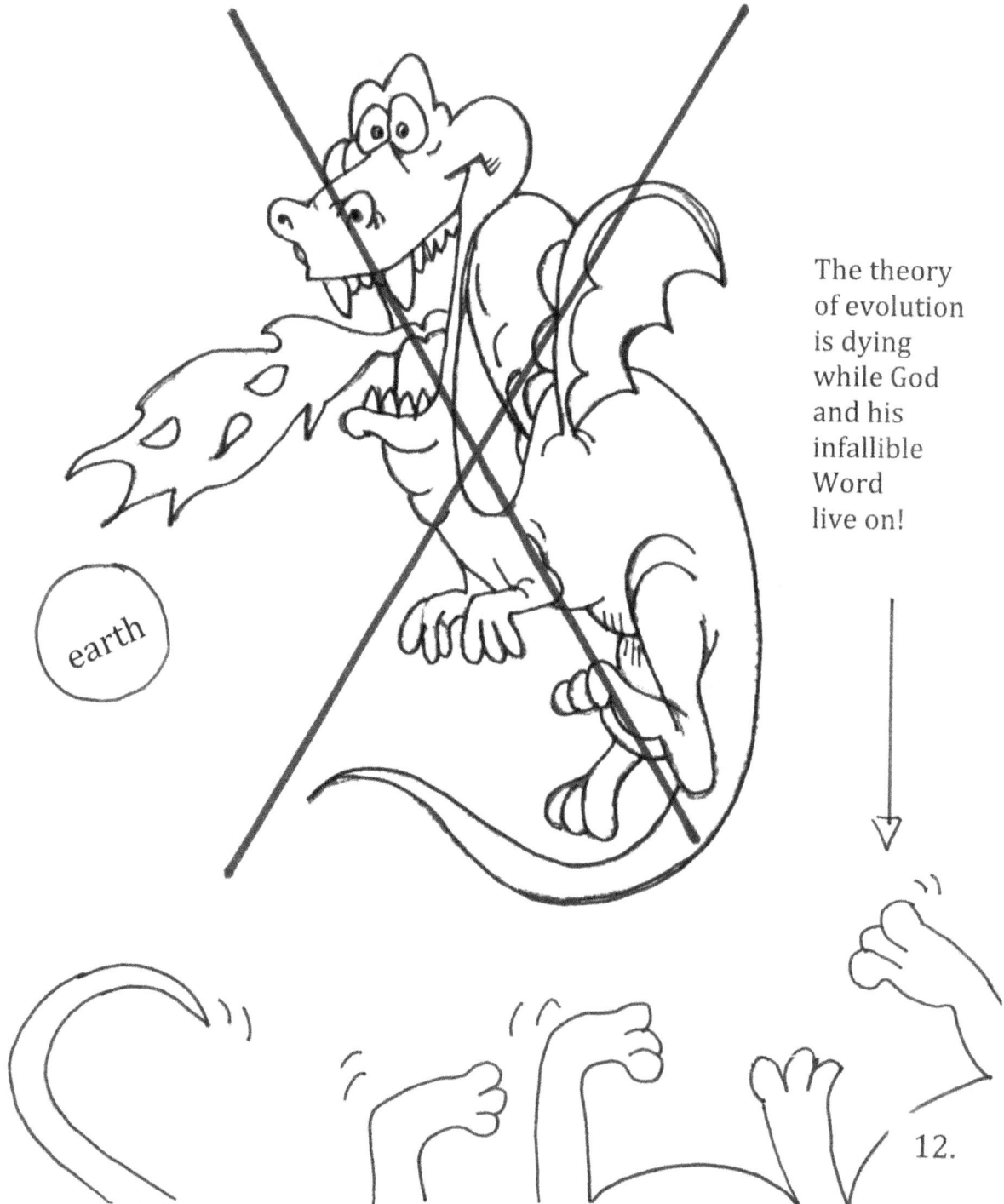

earth

The theory of evolution is dying while God and his infallible Word live on!

Coming up next:
Part 2 is a synopsis of problems and proofs.

Contents of Part 2:

What evolutionists believe

Big Bang Beginnings

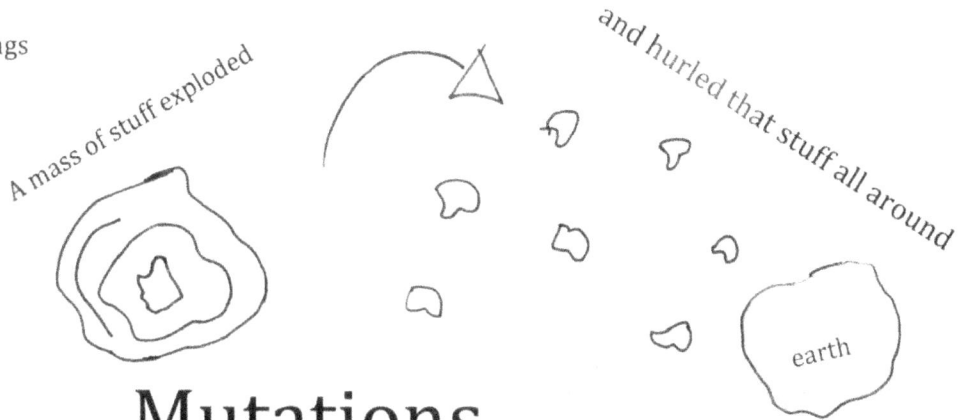

A mass of stuff exploded

and hurled that stuff all around

earth

Mutations

Stuff bumped into other stuff like ocean shores (where did they come from) and caused changes in the stuff that were then passed down to the stuff's children (eventually stuff became apes which then became man.) Wee problem...why didn't all of the apes become men?

ouch!

① Stuff

shoreline

② second generation stuff

"From goo to you by way of the zoo"

Survival of the Fittest

The stronger stuff lasts

The weaker stuff dies

Motivation?
...to eliminate God.

GOD

Why was the big bang theory widely accepted?

Mankind desperately wanted to put forth a godless narrative about how the world began, in order to not have to believe in a God, or worship Him, or become responsible to live in the light of His ways and commands.

Evolution is the biggest false narrative (asserted against all logic and factual evidence) ever to be put forth in the public discourse in the history of man. It has blinded the eyes of billions of people with deception now for two centuries.

What Creationists Believe

1. <u>Codes and info:</u> require a designer
2. <u>Systems:</u> require a designer (not just making the stuff but designing and maintaining the stuff's functions.)
3. <u>Symmetry:</u> requires a designer
4. <u>Symbiosis:</u> requires a designer

To build a house you need an Intelligent Designer

For every house is built by someone, but God is the builder of everything. Hebrews 3:4

17.

Problems with...
the Big Bang

Reverse Spin

Oops ! Venus spins backwards

Water

Evolutionists start with dry matter only....
Where did the vast amount of water come from?

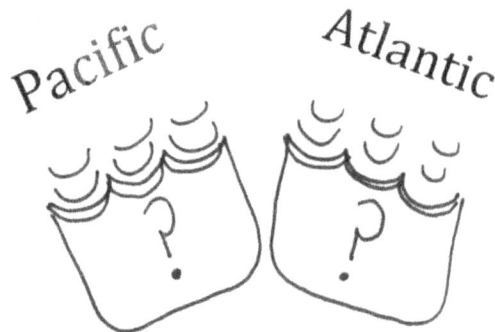

Pacific   Atlantic

Watery verses!

Prov. 8:24
Psalm 33:7
Psalm 148:4
Psalm 136:6

2 Peter 3:6:

"The earth
was formed
out of water
and by water."

Order out of chaos

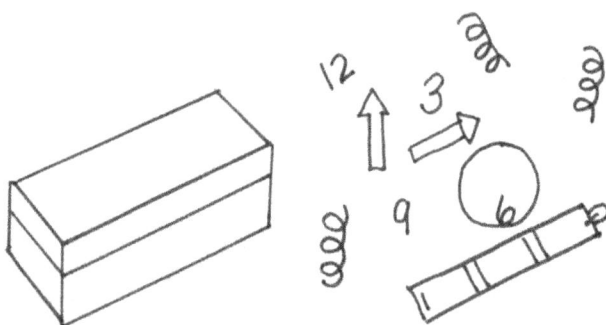

Put the parts of a watch
In a shoebox.
Shake and dump out.
Does a watch appear?

Problems with...
the agents of Evolution

Mutations:

"mutations always mar matter"

good shape

deformed

never reforms

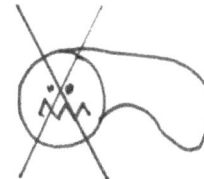

So how can mutations be the agency of taking matter to a higher form?

Thermodynamics

Evolution requires that things evolve to a higher, more well-ordered condition. There is not a single example of this in real life. The second law of thermodynamics states the exact opposite: that things left alone erode and run-down and deteriorate, instead.

a pile of dirt...   erodes over time   19.

Problems with...
Survival of the Fittest

For evolution to succeed, all of the individual parts of an organism have to evolve at the same rate, or it would become so malformed, dysfunctional, and misfit so as NOT to survive! Minor point: mutations always lead to deterioration, not improvement, and never mutate in sync with other mutations in other parts of an organism.

Irreducible complexity

Complex things, be they motors or organs, won't work unless all of their parts work simultaneously. But if each part develops/evolves separately at different rates via mutations, it is impossible to get all the finished object's physiology and function to arrive at their finished state at the same time!   Partial evolution could result in the thing exploding, malfunctioning, or sitting in a stupor, totally defunct.

Mouse trap

Even one part of a mouse trap that doesn't work means the entire mouse trap doesn't work. It needs a base, a spring, and a triggering stick. Miss one of these components and the mouse runs free, even after meeting the trap.

trigger stick

spring

base

20.

Blood clotting

Lots of different enzymes are needed simultaneously to create mending netting

If not all enzymes are present,

there will be no clot !

Lightning bug

TWO chemicals are needed to shoot out the back of a lightning bug simultaneously to create light.

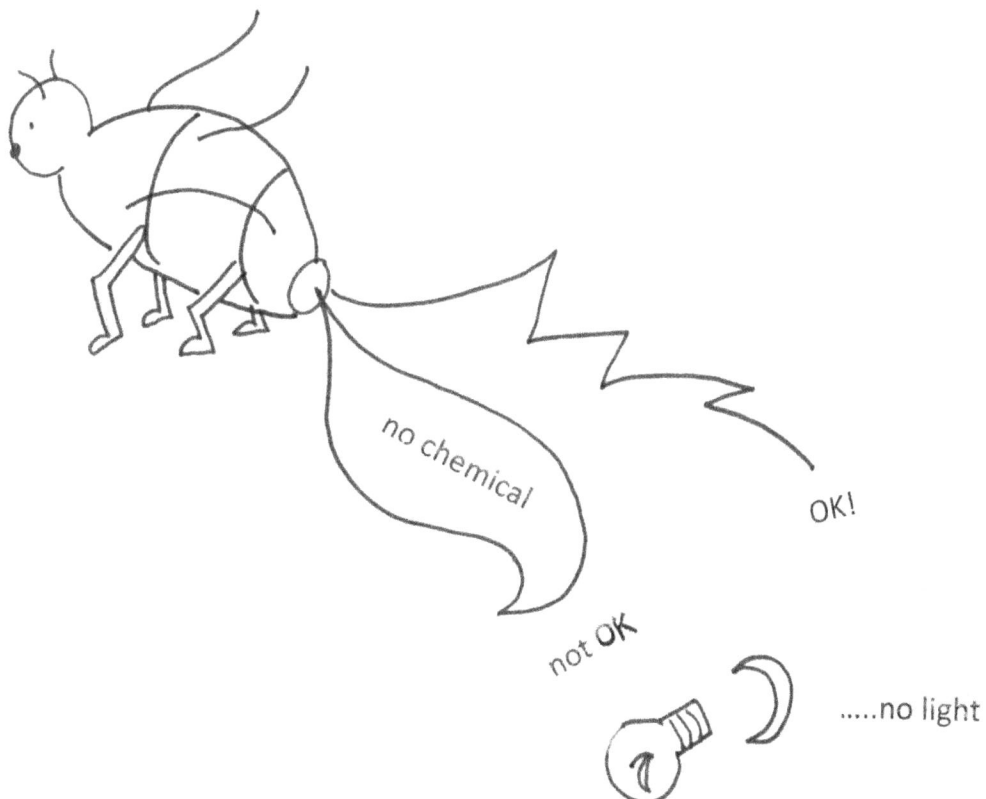

no chemical

OK!

not OK

.....no light

There is no current evidence that animals ever breed across species. No cats mate with monkeys or guppies with toads! The changes that we DO see are only slight changes withIN species.

10X in Genesis 1 God declares that animals will produce only "after their kind," never between species.

A Horse + A Donkey = A Mule
But a mule cannot reproduce. It is sterile.
God ensures there will be "no more!"

Transitional forms don't facilitate survival

How would an alligator survive WHILE its feet were changing into wings?

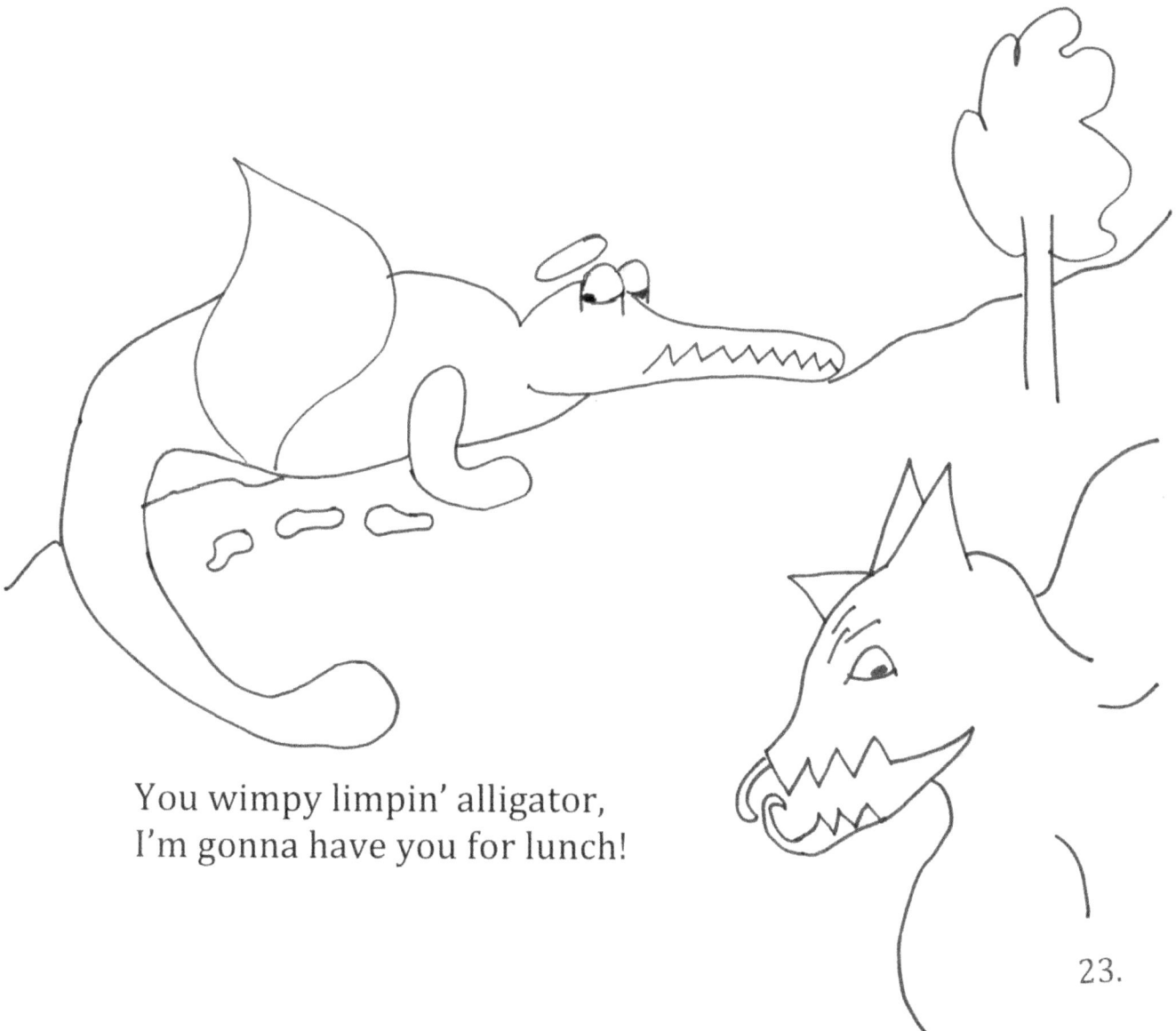

You wimpy limpin' alligator, I'm gonna have you for lunch!

No transitional fossil evidence

later, its highly changed version

original version

fossil silence

Evolutionists don't want any attention drawn to this embarrassing fact.

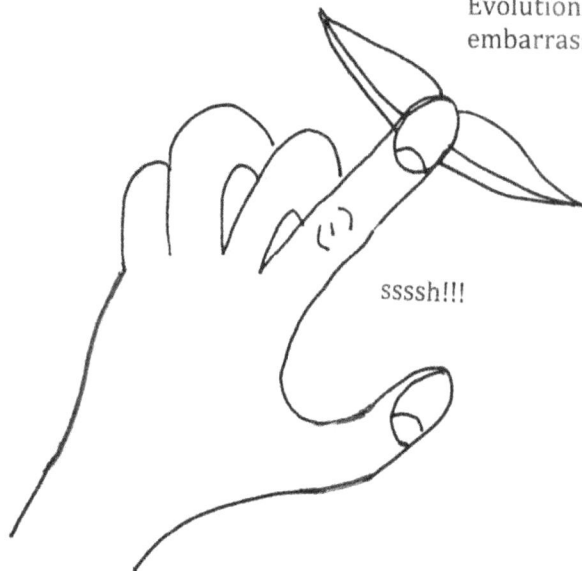

ssssh!!!

A problem with believing that it took millions of years to form the earth

Moon's migration rate

The moon is "leaving town" at the rate of 1.5 inches per year.

Retroactively then...the moon cannot be millions of years old or at the current rate of migration backwards ...it would have been originally located touching the earth!

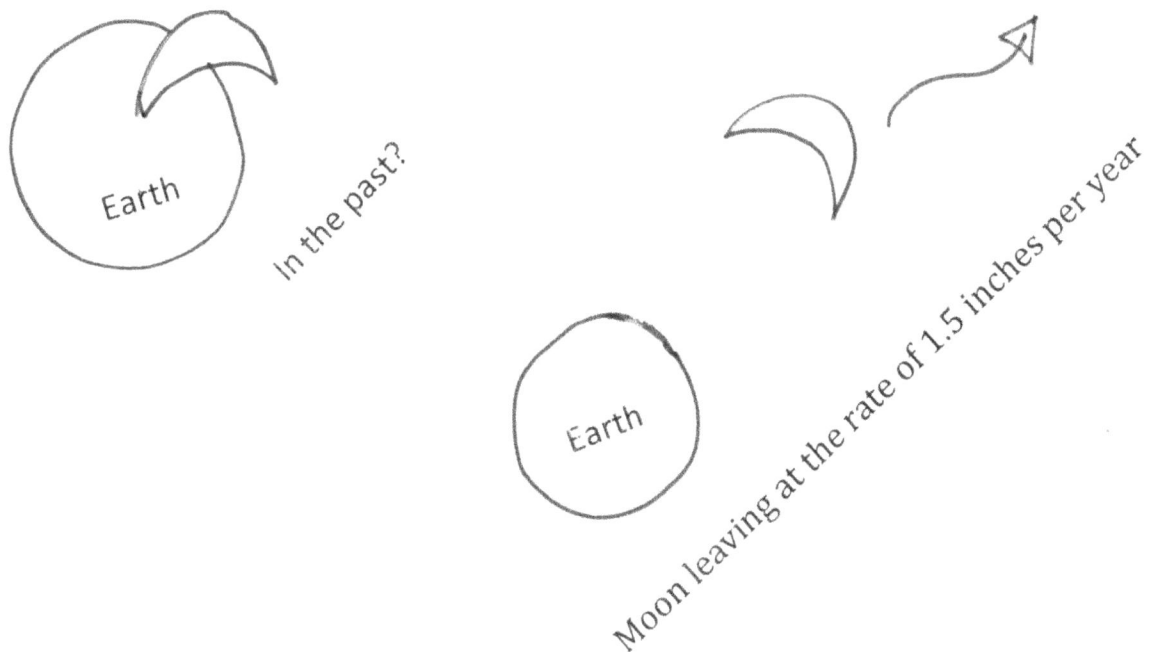

Earth

In the past?

Earth

Moon leaving at the rate of 1.5 inches per year

Sure evidence of a designer

Narrow temperature ranges necessary for human comfort are in fact what we see: our environment is ideally fitted for human survival.

Human comfort zones

The earth's perfect distance from the sun is evidence that a Creator positioned it just so from the beginning.

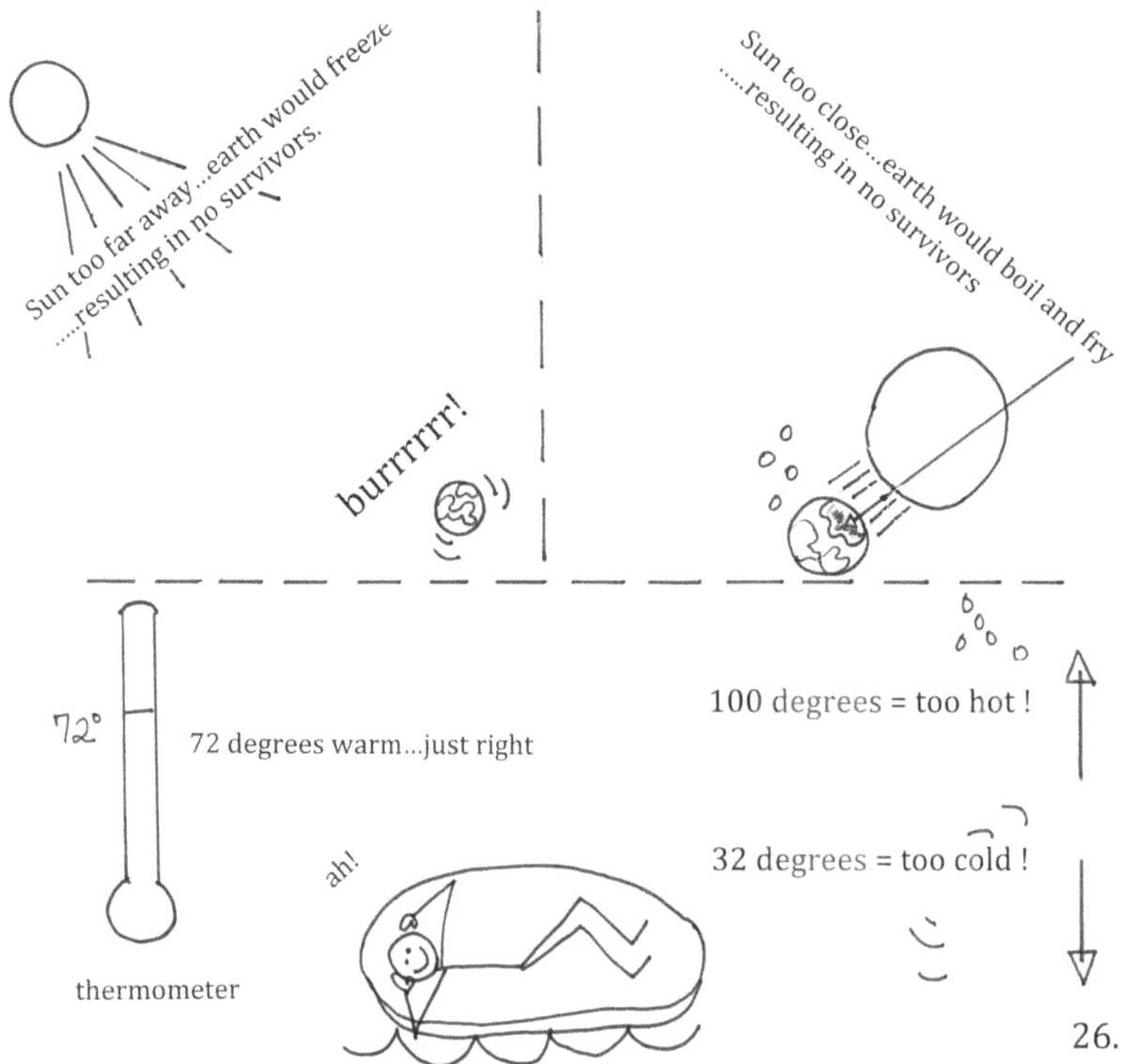

Sun too far away...earth would freeze .....resulting in no survivors.

Sun too close...earth would boil and fry .....resulting in no survivors

burrrrrr!

72°

72 degrees warm...just right

100 degrees = too hot !

32 degrees = too cold !

ah!

thermometer

26.

Proofs for creation

1. Codes

Creation began with information, not with matter.

OICU812   Huh? (hint: say it out loud!)

DNA helix

Hebrew   א ח

Chemistry

Physics

etc.

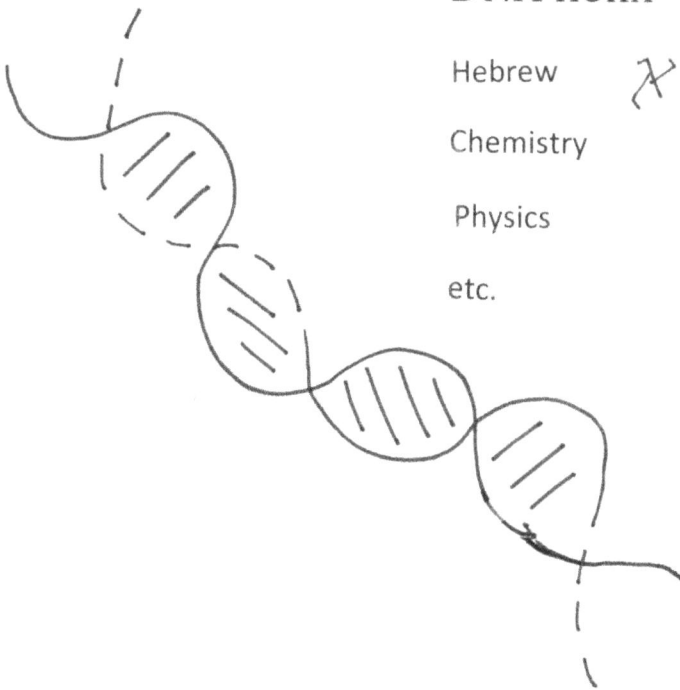

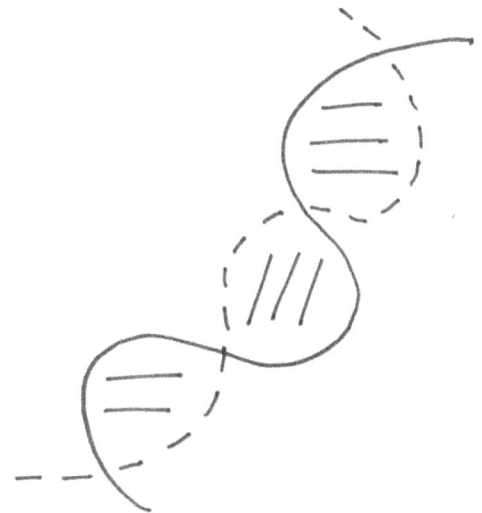

Codes can't amass themselves out of chaos.
Codes have to be designed.

2. Systems:
the organs AND their functions

"He doth create the SEEING eye!" Psalm 94:9

God made a woodpecker's long tongue
<u>AND</u>
a hollow head to store it in!

Tongue rolls up in head like a garden hose

God makes not only the stuff but what the stuff DOES. How the stuff performs had to be invented simultaneously WITH the stuff itself.   For example, it was not enough to make the heart, but the heart has to PERFORM too...someone had to MAKE it ABLE to perform.

heart

lub-dub ⟶

beat-beat

squirt-squirt

3. Symmetry:

eyes

Both eyes are needed for depth perception

ears

Both ears are needed for hearing the direction of sound

thumbs

An opposing thumb is needed to pick up stuff

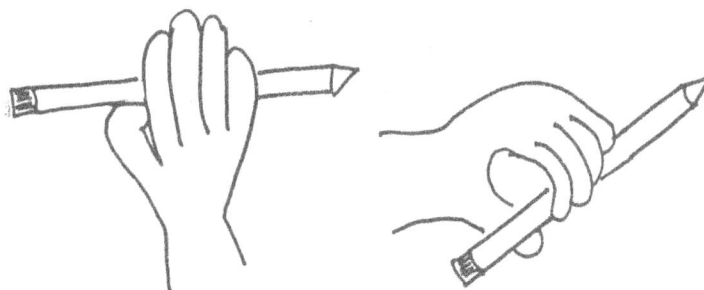

4. Symbiosis: bees and flowers

The flower needs bees to spread pollen to fertilize other flowers.

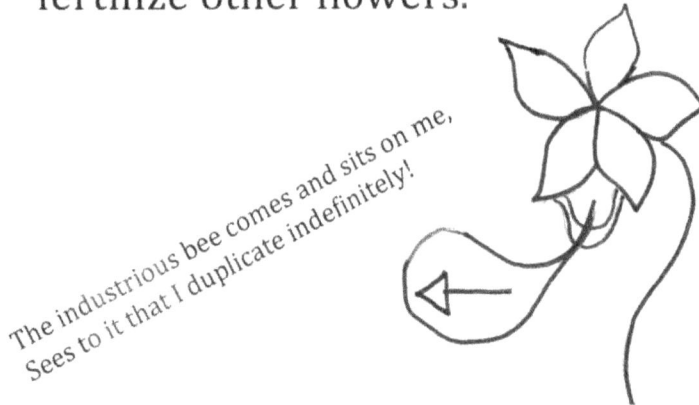

The industrious bee comes and sits on me,
Sees to it that I duplicate indefinitely!

The bee needs flowers to make nectar for fuel.

I've found my flower
she gives me fuel and power.
Sweet nectar I make,
when her pollen I take.

ecosystems

Just like Ben Franklin said of people fighting for the cause of liberty:

"We all hang together or we all hang separately!"

Each ecosystem is co-dependent. It all works together or it all falls apart.

False Assumption: Grand Canyon

A river carved out the Grand Canyon over millions of years. NOT !

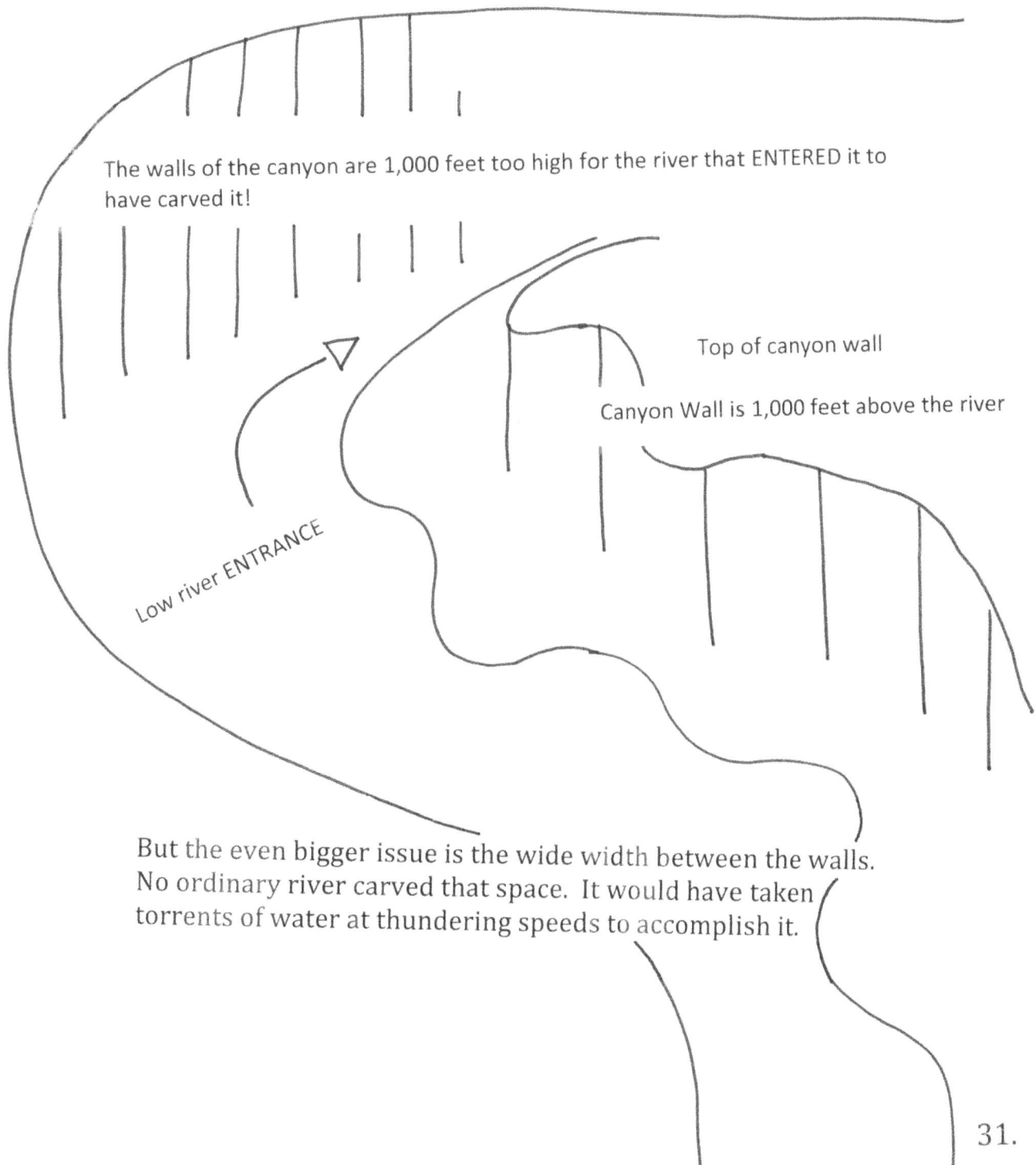

The walls of the canyon are 1,000 feet too high for the river that ENTERED it to have carved it!

Top of canyon wall

Canyon Wall is 1,000 feet above the river

Low river ENTRANCE

But the even bigger issue is the wide width between the walls. No ordinary river carved that space. It would have taken torrents of water at thundering speeds to accomplish it.

31.

False Assumption: Geological Column

Proves evolution. NOT !

Such a column is found nowhere but only in textbooks!

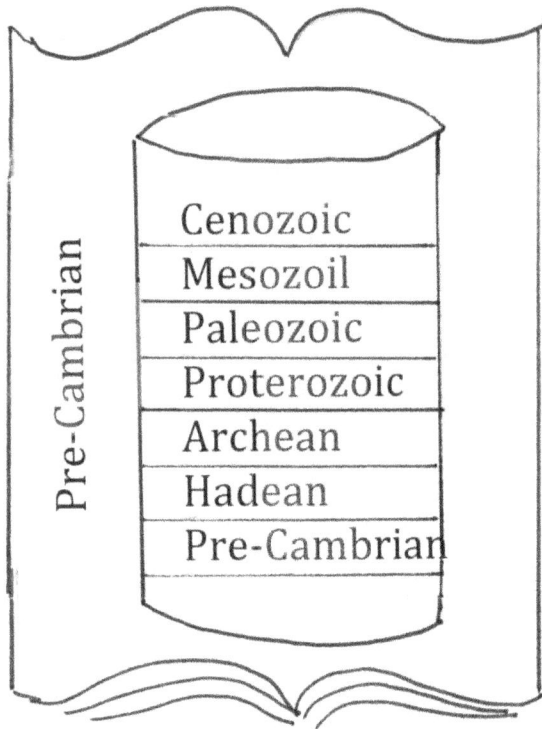

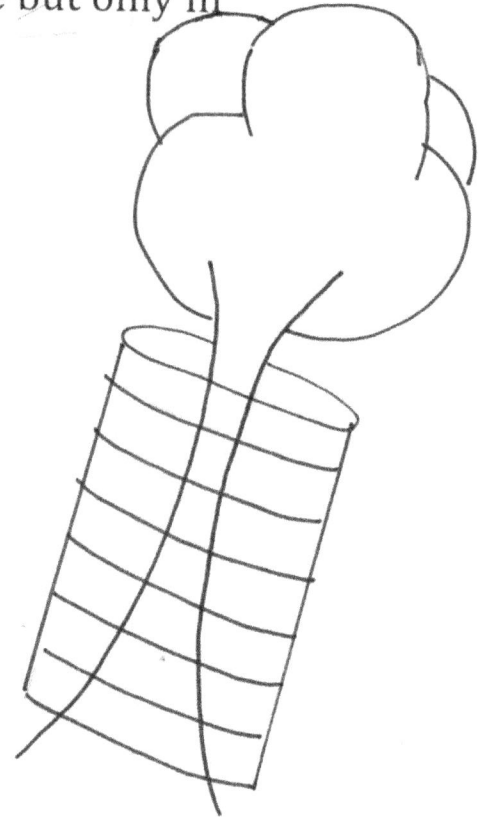

Cenozoic
Mesozoil
Paleozoic
Proterozoic
Archean
Hadean
Pre-Cambrian

Pre-Cambrian

real life

A petrified/fossilized tree found thrust through all layers is evidence of a flood, not millions of years of slow land build up.

False Assumption: Carbon Dating

Carbon dating is reliable, and over time carbon's ½ life evaporation tells the exact date.   NOT !

Carbon dating is UN-reliable. Its ½ life is subject to changes in the weather, heat, pressure and local natural disasters.

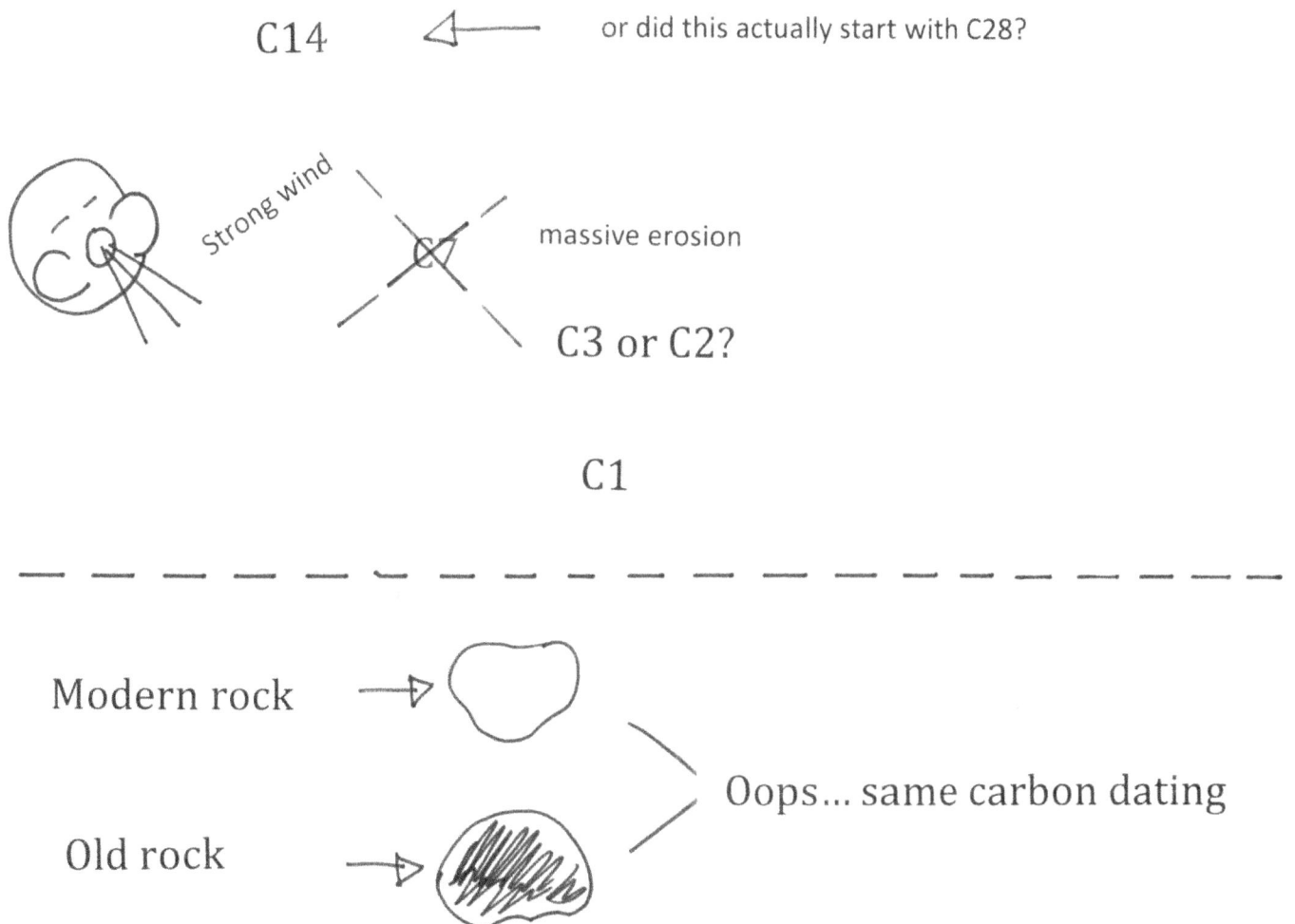

C14   ⟵   or did this actually start with C28?

Strong wind   massive erosion

C3 or C2?

C1

- -

Modern rock →

Old rock →   Oops... same carbon dating

False Assumption:

Caves can only form over millions of years.   NOT!

There are dozens of modern cases of

Rapid Cave Formation

We even have modern examples of caves being created in a few years, months, weeks, or hours.

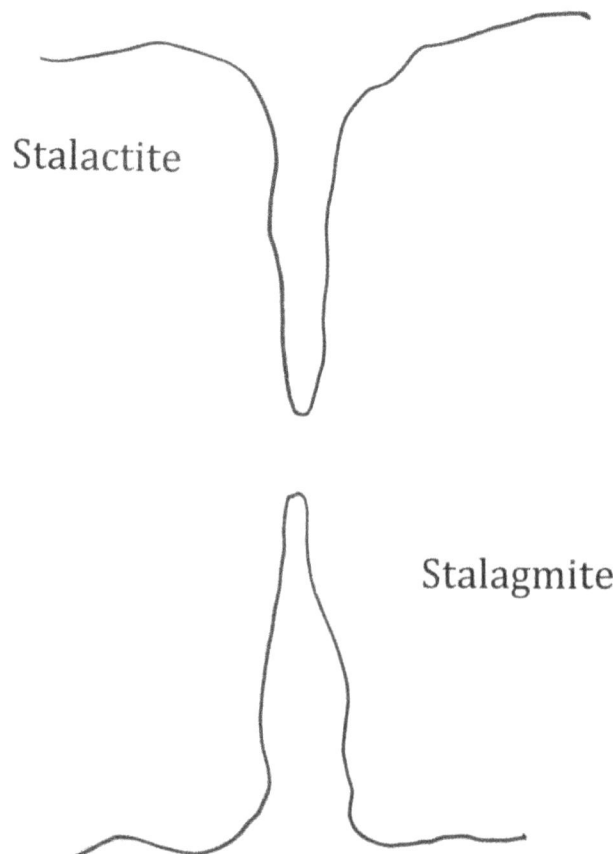

Stalactite

Stalagmite

False Assumption: Vestigial Organs

Some organs are no longer necessary. NOT!
They are unneeded left overs from past forms, and
vanishing by the centuries.   NOT !

Conversely, it has been recently proven that all 3 of these
supposedly unneeded organs are, in fact, needed for immune
function or stability of structure.

Your tonsils are needed

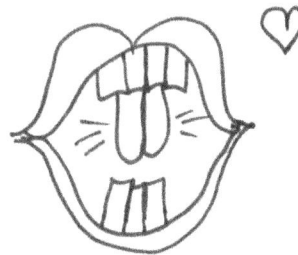

Your appendix is needed

Your tailbone is needed

False Assumption: Dinosaurs

Dinosaurs only lived in antiquity before the creation of man. *Not!*

Dinosaurs were created within the six-day young earth model along with the rest of the reptiles, and lived along-side early man.

"His works have been FINISHED since the creation of the world." (Hebrews 4:3b)

"Look at the Behemoth which I made along with you!" (Job 40:15a)

Man's footprints have been found alongside dinosaurs' footprints in the fossil record.

There are no dinosaurs living today.   Not!

Many sightings of extremely large dinosaur-like reptiles have been found in the Lockness area of Scotland and other South American and African deep waters. Some have washed ashore in modern times, dead, and facilitated even deeper research about them.

Dragons are inventions in fairy tales.   Not!

Dragons are recorded in Job 41: 18-20 as living right along with man.   "Its snorting throws out flashes of light...flames stream from its mouth and sparks of fire shoot out...smoke pours from its nostrils."

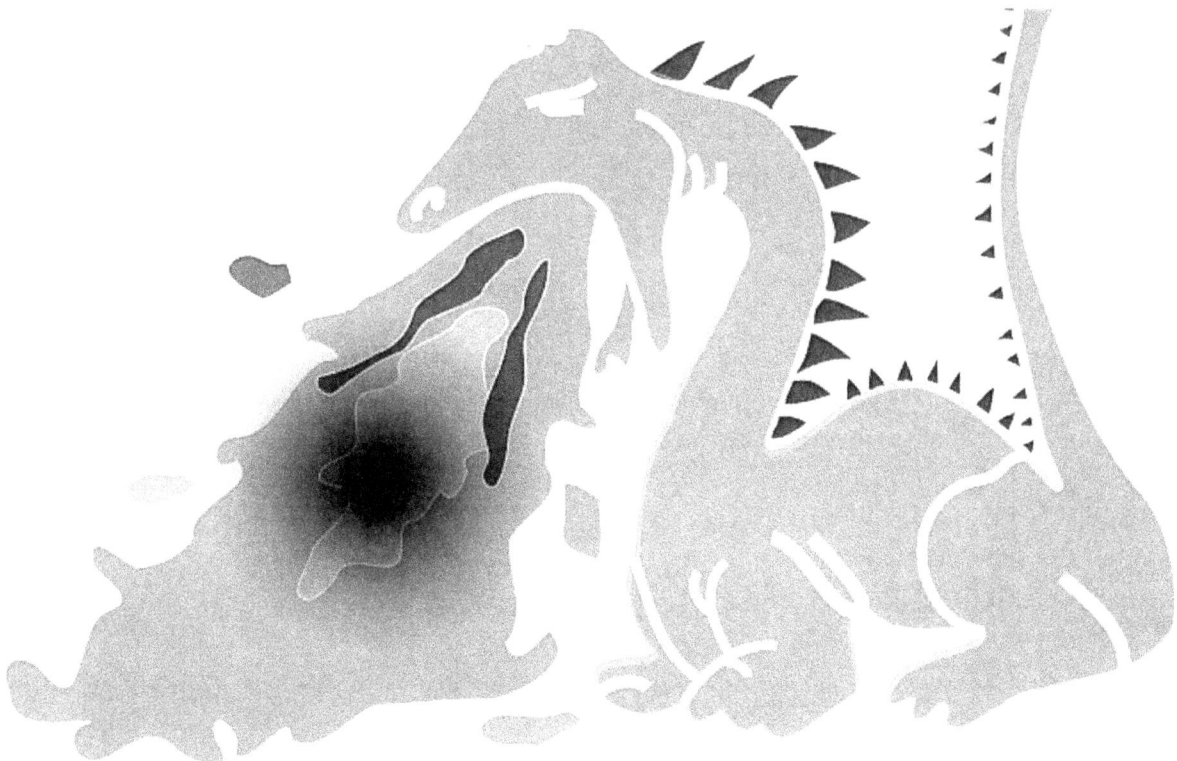

Other Bogus Proofs for Evolution Debunked

Hoax Structures
The Piltdown man and other museum display structures were put forth as proof that they were the missing link. *Truth:* these ape-like skeleton/man models were built around scanty minor random bones, far from the collection needed to construct a model of an entire man.

False Embryo Pictures
False assertion: the developing human embryo goes through various evolutionary animal shapes. *Truth:* the artist who drew these for high school and college texts faked many aspects of the drawings. The embryo stages, in fact, do not look like these.

Differences Only Found Within Species, Never Across Species
The alleged changing evolutionary birds' beaks found on birds in the Galapagos Islands is no evidence for changes across species. Such changes were, in fact, only happening within that species, and often disappeared within the next several generations.

The Best Engineering Savvy Always Employed
The assertion that similar bone structure in jaws and joints of various species indicate a common ancestor is false. Similar bone structure is evidence only of an intelligent designer who used the best structural engineering in all that he created.

Lack of Lab Proof
We have never been able to duplicate in a laboratory the mechanisms evolution claims are needed to create more complex forms. A lab has never been able to spark life from non-life. And a lab has never been able, via natural selection and mutations, to get improved, more complex biological forms as a result. Not once.

Only More Time?
Only applying more time doesn't make impossible things then possible!

Missing in Action
If evolution were true, all things would be in a continual state of change. We would be actually seeing all sorts of transitional forms all around us at all times.

Magic Wand?
How could a solitary cell create enough materials (*stuff*), to fill the galaxy with the multitudes of molecules needed to create everything else?

Darwin's Condition
Motivation?

Pre-set, rigid eagerness to eliminate God

Darwin, the wayward son of a Christian, believing father, rejected his own theory on his deathbed.

His Life. Darwin was born on Feb. 12, 1809, at Shrewsbury, and was educated at the universities of Edinburgh and Cambridge.
His father wanted him to become a clergyman, but ...

Brown Bros.
Charles R. Darwin

His unenlightened science

Darwin thought the cell a blob of stuff

Oops...microscopes revealed a mass of information inside the cell similar to computer codes, along with biological apparatus to carry out what the information said to do !

39.

Ideas Have Consequences

Holocaust

Evolution's core belief of "survival of only the fittest" lead to the belief that the Germans were a superior race fit to live but Jews were not fit.

The unfit must be extinguished, including any unwanted, unborn babies via abortion. We can ultimately do all the evil we'd like.

Philosophical incongruity

Without God....

We can fashion life to be any way that we ourselves design.
We can assert any reality and claim that it is true: i.e.
"I am a woman even though I have male chromosomes."
"I am black, even though I appear white as an albino."

Implications

There is no right and wrong.
There is no judgment
There is no objective true reality .

A Comparison of the Two Theories of the Origin of Life

Evolution	Intelligent Design
Author/theorist: Charles Darwin	Creator: God, as detailed in the Bible
Everything originated from rocks and dirt	Everything originated from water (2 Peter 3:6)
It was explosion-based (big bang of matter)	It was information-based (complex DNA codes)
It developed over millions of years	It began in six days—only 6,000 years ago
One common ancestor: a one-cell organism	Different kinds (species) existed from the start
One-cell (simple)	Each cell is incredibly complex
Survival of the fittest	Survival of all, simultaneously
Obsolete organs (tonsils and appendix are deemed useless, unnecessary)	Each organ is important (tonsils and appendix are vital elements of immune system function)
Ignores the empirical evidence that mutations nearly always have a negative consequence	Mutations are always destructive, they are never improvements
Ignores the fact that some things don't work unless each part is working (e.g., the mousetrap)	Irreducible complexity within organs is demonstrable (e.g., blood only clots when all of the chemistry triggers it at once)
Ignores the necessity of synthesis and symbiosis for mutually dependent eco-systems to function	Each eco-system had to develop at the same time as its individual parts
Views the possibility of a horse and a donkey producing offspring as an evolutionary triumph	A horse and a donkey produce a mule, which is sterile (cannot reproduce)
Relies on the carbon dating of fossils	Carbon dating has been found to be unreliable.
Claims there are transitional fossil forms	No transitional forms between species are found in the fossil record

So take your pick...

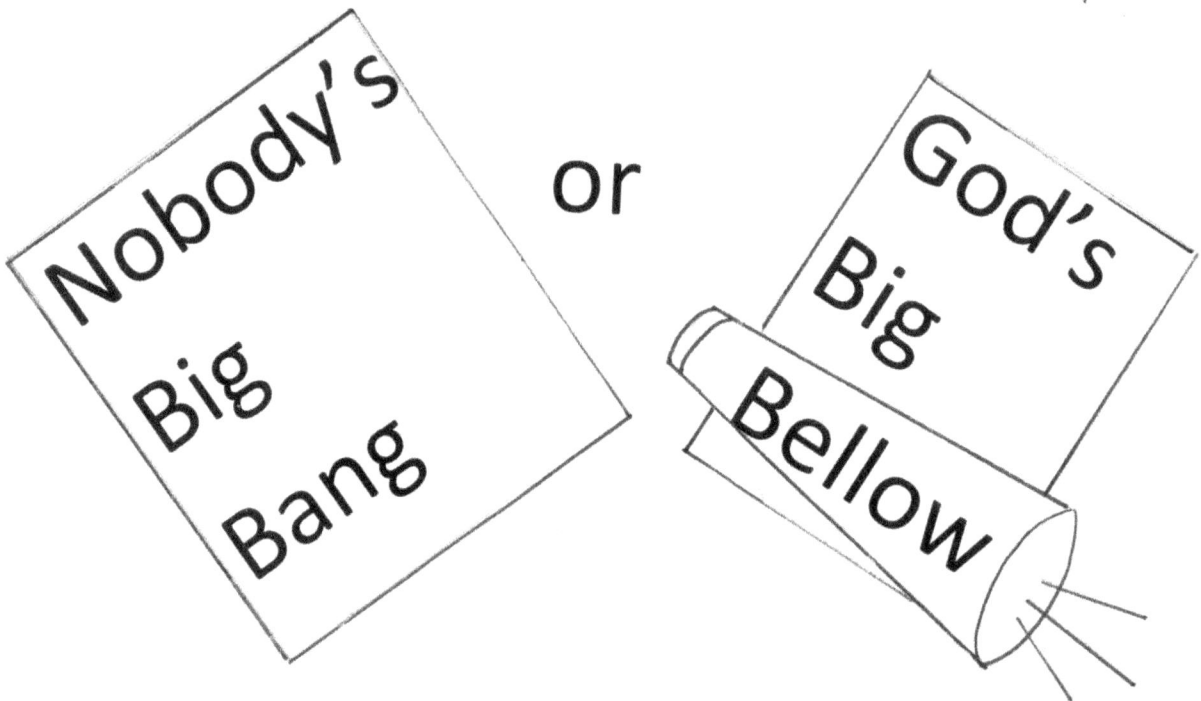

Nobody's Big Bang

or

God's Big Bellow

How wonderfully rational it is to ...

Vote for this ! ✓

God said it and "BAM"! It happened!

A Biology Graduate's Conclusion about Evolution's Viability

Contrary to what you might expect, majoring in biology at a secular school firmly demonstrated to me the complete impossibility of the theory of evolution. The more you understand about the underlying complexity of life, especially the biochemical pathways required for even the simplest of functions, the more it becomes clear that evolution is a series of just-so stories completely bereft of a plausible mechanism by which it could occur.

The more the teachers would say "isn't it amazing that evolution did this," or worse tried to downplay the amazingly intricate nature of some facet of creation, the more obviously it would become that they had no real explanation. Evolution was just a catchall, throw millions of years at something, and somehow the impossible happens.

Everything we observe on the other hand shows the exact opposite. Life does not arise from non-life. Complex organisms don't change into new forms with new functions, they lose function, inherit greater genetic burden, and, if anything, devolve in order to adapt to a fallen world.

Answers in Genesis, and the Institute for Creation Research, among others, were invaluable resources during my time there in the Darwinian den. My professors were always quite surprised to find out I was a "catastrophist" as one called me, but they really didn't have any good answers when challenged, just repeated their standard boilerplate.

There were other students, too, who challenged the consensus. When one of our biology teachers decided to have a debate within the class, our group was able to get him to concede that origins is outside the scope of science, being inherently unobservable, which is a fatal chink in the materialist position. If not everything can be known materially, then you must concede that it is possible that there is more than merely the material.

So all in all, the more I learned about actual science, as opposed to the materialist religion they try to substitute in its place, the more fervent I became in my belief. Science is merely a method by which one seeks to find and apply the truth. Absolute truth exists as a necessary presupposition to science's function.

S.P., Kansas

Theistic Evolution
A Nice Compromise?

(A theistic evolutionist is someone who believes in both the Bible and evolution)

But it is impossible to believe that it took millions of years to form the earth and still believe in the Bible. Because if you believe the Bible, you can't have death occurring before sin. Sin came through Adam, therefore no animals died before the creation of Adam.

> In Adam we all die. I Cor. 15:22
> Just as sin came into the world through ONE man, Adam... Romans 5:12
> (This second verse specifically addresses those who erroneously think Adam is a term used to describe all of mankind.)

Resources

Websites:
Answers in Genesis (AIG)   answersingenesis.org
Creation Research Institute (CRI)   cri.org/discover

YouTube Resources

Evolution is impossible, not just unlikely.

Professor Exposes Impossibilities of Evolution
(Professor Jonathan Wells of UC Berkley)

Honest Science Proves Evolution Impossible.

David Berlinski Explains Problems with Evolution

100 Reasons Why Evolution is Stupid

The Failure of Darwin's Theory

5 Scientific Facts Prove the Theory of Evolution is False

The Biblical View of Creation
(some heavenly thoughts)

Romans 4:17:
"He calls into existence the things that do not exist."

God is the only "being" who can make something out of nothing. All artists henceforth recombine what already exists; they never are able to materialize the raw, original matter. Mankind's derivative projects of "created beauty" are limited only to new combinations. Humans' art can be beautiful, but let us never forget that it is of an altogether different quality than God's.

Colossians 1:15-17:

"The Son is the image [face] of the invisible God, the firstborn over all creation."

He who was co-existent with God, nestled inside of God, was God Himself, and underwent an expulsion/birth at some point as the supreme corporeal, tangible expression of the entire Spirit-Father. The Son is the visible FULLNESS of God (Colossians 1:19). He became the seeable, touchable face of God. This birthing of the tangible God Himself into Christ was a necessary prerequisite for there to even BE a subsequent, further, temporal Creation.

(Colossians 1:15-17, continued)
"For in Him all things were created: things in heaven and on earth; visible and invisible, whether thrones or powers or rulers or authorities;"

In and through Christ, the world materialized and took on visible embodiment. First He created the offices of administration, thrones, rulers and authorities, and then He filled them.

The Lord spawned invisible metaphysical matters like morality, too. Love was not made out of atoms, for example; but it did, nonetheless, have to be fashioned to express itself "thus-and-so". And all of this is to say nothing of that further activity brewing in the rest of the mind of God, where He is doing all manner of who-knows-what," elsewhere, outside of time and outside of earth.

(Colossians 1:15-17, continued)
"All things have been created through Him…"

The godhead's corporeal begetting of the Christ was absolutely necessary to expedite all the rest of Creation. Christ was integral to the whole. He was not only the supreme catalyst of all the subsequent creation, He was the very agent of it.

(Colossians 1:15-17, continued)
"And FOR Him."

God designed (and purposed) to extend His glory through the huge and gallant work of Creation. Apparently He needed the creation of the world through which all things physical and metaphysical would be reconciled TO His Son—brought to Him--placed beneath His kingly feet, to the end that His Son would pre-eminently preside over its future consummation, for some glorious high purpose.

All of this story was hidden from our understanding (no doubt for divine reasons), generated in the pre-existent eternal councils of God. "…this mystery which for ages past was kept hidden in God, WHO CREATED ALL THINGS …according to His eternal purpose" (Ephesians 2:9-11). So, apparently, something of great importance had to be accomplished both in a creation through Christ and by a redemption through Christ. The Lord, in both ways, was the divine funnel of the extension of God.

(Colossians 1:15-17, continued)
"He is before all things..."

Evolutionists insist that no god is needed for Creation to have taken place. Conversely, what the Bible tries to tell us again and again, in every way possible, through multiple verses found all throughout Scripture, is that Christ is an integral part of Creation. Without Christ there could have been no Creation. He was created FOR creation by becoming the firstborn OF creation. He is the a priori source of creation and is equally the condition OF creation.

(Colossians 1:15-17, continued)
"...and in Him all things hold together."

If Christ would ever cease to think of Creation, it (the created) would disintegrate. The world holds together by His divine energy and wisdom—through His divine workings of biology, chemistry and physics and through His judicious ongoing administration of righteousness. His eternal being continually acts to hold the cosmos together. Not only is His divine activity the glue of creation, His mere existence, even without that activity, is its very sustenance.

Romans 1:20:
"For since the creation of the world God's invisible qualities..."

....were not understood or seen before the Creation. They existed in abstraction, in etherealness.

(Romans 1:20, continued)

"His eternal power and divine nature have been clearly seen, being understood from what HAS BEEN MADE."

For some reason it was necessary to extend the knowledge of God by viewing material matter and moral law.

John 1:2:
"Through Him all things were made; without Him nothing was made that has been made."

Nothing was made outside of Christ's having made it—Christ commanding it—Christ WILLING it—presiding over it. There were no add-ons, no chemical reactions going on outside of His purview, apart from His extraordinary instigation, wise administration, and holy jurisprudence. In other words, there were no sideshows or derivative acts of origin going on. The initial six-day Creation was all orchestrated and created THEN, and was finished--not to be created out of nothing ever again, inside of time.

A crucial choice
What we decide to embrace as our personal creation model is the difference between believing in a creation of chaotic randomness vs. a creation of mind-blowing specificity. The Creation that Christ produced was brought forth in infinite detail and is sustained by His infinite micromanagement. "He slumbers not nor sleeps" (Psalm 121:4). What a comfort it is to know that there is a Person at the center of our existence, and not some piece of rock.

Seen in this light, the theory of evolution is rankly evil, a vain argument lifted up against God's supreme eternal purpose for creation and for His Son.

Praise Him for his acts of power; praise Him for his surpassing greatness, for at His command the worlds were created.

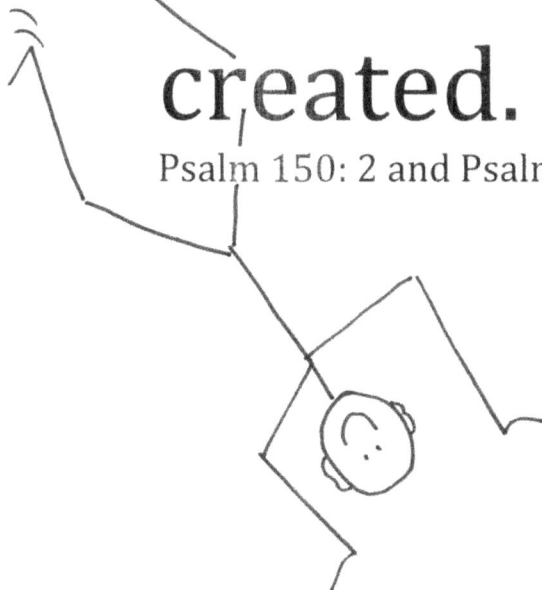

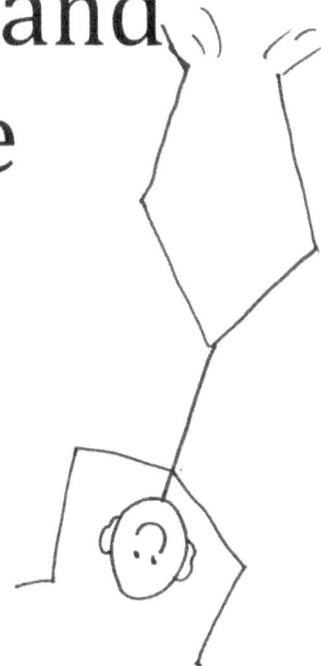

Psalm 150: 2 and Psalm 148: 5

Why Creation by God is Inevitable and Evolution is Impossible

It just may be that some who are taking up the debate of Evolution vs. Creation are leaving the discussion far too early, because they focused upon the HOW of creation and left the WHY of creation miserably underexplored.

The HOW has involved three distinct possibilities:

One:
Creation was a random cosmic explosion with no meaning, followed by millions of years of evolution.

Two:
God started the whole affair, with a little bit of meaning, not trying to say anything particular about Himself, nor show any significant difference between man and beast (since God is totally OK with letting man evolve from beast), and then let evolution complete it over millions of year. (This is called Theistic Evolution. Most believers park here with this assumption.)

Three:
God initially zapped all origins into existence with no help from natural causes, for a power-packed display to say something about Himself and something about man.

And there we sit with those three options, often no further along with any conclusion than when we started. But, as we shall soon see, the WHY of creation may well lead us straight to its HOW.

Let's look at an earthly example: Building any earthly building must be preceded by its WHY. We determine beforehand how the building will be used, in order to know how to build it. Will our building be a little eight-foot-wide popcorn shop in a crowded retail space on a narrow street of downtown Chicago—or will it fill the area of three huge parking lots to house a large manufacturing business in rural South Dakota?

Now let's extrapolate that to the heavenly business of building a world. What was its purpose? We are told in the Scriptures in two places that the purpose of creation was to define or express God in some extraordinary way.

Romans 1:20: "...for since the creation of the world God's invisible qualities – His eternal power and divine nature – have been clearly SEEN, being understood from WHAT HAS BEEN MADE."

Psalm 19:1: "The heavens declare the glory of God."

So, if that is the purpose, then a creation via a godless random explosion followed by a long, meandering, godless slow natural process of some sort would be utterly meaningless to that purpose. If creation was intended to show the glory of God but we must begin with godlessness, and continue with chaos, this is absurd. Why involve God with a creation at all?

But let's imagine that there IS a God and that it IS the case that God wanted to show something of His character and nature through His act of creation. Easy enough for an Almighty God to perform in any way He so chooses, correct? He COULD have used a big bang (though chaos would have been a strange modus operandus for displaying His glory and His perfections), but let's say that we allow for such a point. That would be feasible IF that were all there was to it.

But it is very curious, indeed, to find that that was NOT all there was to it. The Scriptures further tell us that God limited/restrained the entire act of creation to be carried out only via His Son! It was to be done carefully and explicitly, by fiat, His Son's spoken word, "by the breath of his mouth" (Psalm 33:6). From the get-go, that explanation of things would rule out a random explosion. For how could a long drawn out affair be spoken? How could millions of years of words be viewed as an explosion? Such an oration would make quite the drawling sentence!

This is one problem. But here is yet another. Apparently it was not enough for God the Father to dream up creation within His own genius and then immediately execute it by Himself. Instead, He mandates and insists that the act be carried out BY His Son, THROUGH His Son, and FOR His Son of which His Son was the firstborn OF all creation. If Christ is the firstborn of creation He, Himself, certainly wasn't born of a chaotic explosion! On the contrary, He was meticulously fashioned and was, in some unimaginable way, emitted or "born" out of God Himself.

But let us return to the WORLD'S creation. Note the four prepositions inextricably bound up with the creative act:**"by"**, **"though"**, **"for"** and **"of"** Christ. If this is the case, how then could we have a creation at all, if there is no Christ? It appears that, without the Son, God didn't dare touch any thought of a creation for an entire eternity past.

It was only after the co-existent Son was extruded from God Himself, to express the very face of God, i.e., He became the fullness of the godhead bodily (Colossians 1:19) that the further idea of the Creation was finally drawn forward to the frontal lobe of the triune God-- to be considered within His own divine counsel.

A secular, atheist evolutionist theory of creation requires a godless beginning. One must ask whether this could have even been possible (there had to have been a first cause, a maker, at least of the first matter), but let's concede the point and pretend that it IS possible to have the first matter beginning without God; one must ask, "To what end?" What would be the purpose of such a bizarre, aimless, happenstance creation of endless, meaningless stuff?

In an insane asylum an insane person bangs his head against all four walls, deliberately making this his chosen career, but to what end? Some famous atheists ended their lives in suicide because meaninglessness simply could not sustain them. A creation without either a self-determining or God-determining meaning is a black hole. It simply collapses within itself. Creation must have a meaning or it will not endure; it will eventually disintegrate or be bombarded and crushed, with other wild random acts coming out of nowhere, for no purpose. For, after all, if a big bang happened once it could certainly happen again, out of the blue, utterly demolishing the first, at any moment.

So we see that the meaning of creation powerfully dictates the narrative one must have to describe its beginning, retroactively.

According to Scripture, for the believer there can be no creation if there is no Christ. The two are inextricably bound up with each other. Creation hangs upon Christ, or there is no creation.

To believe in Theistic evolution is a non sequitor--an impossibility. A Theos would have no use for a creation at all, of any kind, let alone one involving a long random evolutionary process, if that creation comes about without Himself, or without His Son or without any meaning--all of which defy Scripture:

Without Himself?
Genesis 1:1: "In the beginning GOD created the heavens and the earth."

Without Christ?
John 1:3: "Through Him [Christ] all things were made. Without Him, nothing was made that was made."

Without meaning?
Romans 1:20: "...His divine nature has been clearly seen through what has been made."

The conclusion of the matter:

Let's return to our original premise that the WHY of creation could lend some strong insight into the HOW of creation. If you were Almighty God (i.e., that was your very essence) and your purpose was to display your glory through creating a world, would you take millions of years to "get it there?" Would you cause all your trees and all your animals to suffer through millions of years of bewildering cross-species changes, that never arrive, as one of your highest means to express your glory? Would you trap and handicap everything in that creation under deforming biological mutations as your agent to improve things? Would you really generate economics and a justice system from a rock and a bang? Could you do no better? "Is the Lord's arm too short?" (Isaiah 59:1 and Numbers 11:23). It is a ludicrous question.

Let us demolish arguments and every vain imagination
that sets itself up against the knowledge of God.
2 Cor. 10:5

The Special Creation vs. Evolution debate

Examined one more time, from a different angle.
This time: via questions galore .

Contents of Part 3:

Evolution is more than a scientific theory; it is part of a structure of religious belief that, at its essence, removes the Creator from His creation. Save yourself hours of wading through technical material. Here's a summary of some of the most powerful philosophical and conceptual arguments against evolution. Pass them on to your children. Get the job done in 15 minutes!

I. The big bang theory

What evolutionists believe:

That all of creation started with a big bang, millions of years ago. A subsequent swirling mass threw out chunks of matter that would continue in similar orbits; obviously they would be made of similar materials.

What's the matter with that view?

1.   To believe that first there was nothing and then that "nothing" exploded is irrational.

2.   Explosions cause chaos, never order. What does a building look like after it has been bombed? An explosion in a match factory doesn't produce order...or human comforts of any kind.

3.   It defies the second law of thermodynamics. Everything wears out and runs down over time. All about us we see that food rots, cars break down, ill health emerges in old age, etc.

4.   Cosmologists observe that stars are burning up their fuel, not making new, and that some stars have died altogether.

5.   If the big bang theory were true, how come some of the planets and moons spin backwards? Pluto and Venus rotate backwards. Neptune, Saturn and Jupiter all have moons that rotate in opposite directions. Uranus is tipped onto its side and rotates like a wheel. We challenge anyone to shoot a BB gun and have one of those pellets spin backwards. It goes against the FIXED scientific laws of motion.

6.   The moon has a completely separate orbit from Earth and is not made out of the same material as Earth. It couldn't have originated as a chunk of matter flying off the Earth.

7.   Humans can only live in a narrow comfort zone. There is a delicate balance evident in everything on Earth in order to sustain the fragile environment so life can exist here. In other words, if Earth were five percent closer to the Sun, the oceans would boil. If we were one percent further from the Sun, the oceans would freeze. After the "big bang", how come the sun came to rest at exactly the perfect spot for human comfort on our planet? The universe was not started with a BANG, but with a WORD. The Creator *set* the Sun in place; He didn't *roll* it there.

8.   If there were more oxygen in the ozone layer around Earth, we'd burn; if less oxygen, we'd suffocate. If there were no water here there would be no life of any kind. If the moon's gravitational pull were even a small percent stronger it would drown Planet Earth with tidal waves twice a day!

9.   Because gases dissipate rapidly, we shouldn't expect the most distant planets to have any. Yet Saturn and Jupiter are gaseous. And, why do four planets have rings and the others don't, if they all were spun off of the same big bang?

10. The Sun is steadily backing away from the Earth, and is losing weight every day. How big would the Sun have had to be a million years ago?! If the Earth and the

Sun were both billions of years old, they would have been touching in the beginning and as a consequence the Earth would have been burned up.

11. If the Earth is really billions of years old, why are there no fossil records before the Cambrian period? Why is the lunar dust only one-half inch thick? (If the moon were billions of years old, its lunar dust would have been so thick, no space rocket could conceivably have landed on it.) We can measure how old or young the Earth is by measuring the rates of present change and then backing that up to see if the thing could still have existed past 6,000 years ago at the present rate of change. Niagara Falls has moved south by seven miles. The moon is growing farther and farther from the Earth. At its present rate of movement, it would have been touching the Earth way before a million years ago. The rate that the Mississippi River delta grows shows that it could not have existed such a long time ago.

12. How come we see a tree jammed between four strata of rock ...the same young tree fossilized standing up, between four different AGES of rock formations supposedly laid down over millions of years? (How 'bout, instead, it jammed, i.e., wedged itself between four layers of rock that heaved when torrents of roaring water swept through during the worldwide flood?)

II. The spontaneous generation theory

Evolutionists believe that life can spring from non-living chemicals--from dead matter. This is called spontaneous generation.

What's the matter with this view?

No scientist has ever been able to produce life from chemicals in any laboratory. A scientist named Stanley Miller tried to get amino acids by applying a spark to gas and all he got was a destructive tar.

III. The theory of evolution from a single cell to complexity

What evolutionists believe:

Even if spontaneous generation is impossible, evolutionists hang the rest of their whole theory on the belief that life evolved from single-celled organisms into complex plants and animals. Their end conclusion is that man evolved from an ape.

For this to occur, two things must have happened:

1. Mutations: an error or alteration appears in a gene and is then passed on.
2. Natural selection: only the strongest and most adaptive plants and animals will survive. This is called the "Survival of the Fittest."

What's the matter with this view?

1. Mutations are almost always destructive. They lead to structural impairment, genetic disease, and death. The ratio of harmful to beneficial mutation is 10,000 to 1. How could such a change benefit any organism?
2. A mutation never introduces a new level of complexity.
3. How would evolving creatures ever survive the "survival of the FITTEST test" to even stay around long enough to mutate further? An evolving creature in its unfinished/middle state would be weak and easy prey for a predator. A mouse with three legs would be eaten by the first cat. If an alligator were limping along the highway with a heavy half-wing hanging off his side, how could he now either scurry out of danger on his old overburdened legs, or fly with such poor newly developing aviation equipment?
4. In order for a mousetrap to work, ALL of its parts must work. The trap doesn't work anywhere along the assembly line except at the end when all of its parts form a complete unit. This is called *the law of irreducible complexity*. Here are several additional examples of this law in action: if the two gases that shoot out from the rear end of a lightning bug had evolved separately, the bug could have exploded. Bats depend on a complex sonar system to be able to move about; without a part of it they would die. Several systems are

involved in the function of the human eye, without which there is no sight at all. The blood-clotting chemistry in humans has several chemistries happening at once; all of it is needed or none of it works. The woodpecker has a film that closes over his eye to keep wood chips from flying into his eye and hurting it as he pecks away. The dragonfly has two different sets of wings, with different controls on the front wings than on the back wings. The chance of all the companion systems evolving at a compatible pace with each other is zip, zilch.

5. Evolution presupposes the evolution of all the inter-related EXTERNAL SYSTEMS that are dependent upon it....i.e. eco-systems. How come this isn't in total disarray? Vitamins fighting with minerals instead of working synergistically? Plants emitting CO_2 instead of oxygen, suffocating us all? CO_2 made by mutations anyway? Where did IT come from?

6. No transitional forms have ever been found in the fossil record...or in the real world. We don't see anything in an evolving state...becoming something else...either slightly beginning or almost there...and nothing that is dead-center in the middle of its evolutionary process. If primates evolved into humans, why don't we see a half ape/ half human in the MAKING, anywhere on Earth?

7. You can cross a horse and a donkey and get a mule. But the mule is sterile and cannot reproduce. This transitional creature simply cannot produce another transitional creature.

8. Today, we are losing species. Species are becoming extinct. We are not seeing the formation of new species anywhere today.

9. If a bird evolved from a fish, whom would the first bird mate with? How can male and female ANYTHINGS evolve at the same rate in order to PRO-CREATE, in order to have MORE such creatures to do further mutations ON, if mutations happen slowly one at a time, cell by cell, RANDOMLY, over millions of years?

10. Have modern microscopes ever observed a single case of mutations happening in PAIRS? If not, how do we get two matching eyes? or ears? And how about the difficulties of symmetry evolving? How could perfect symmetry evolve? A human's skeleton is the same on the right side as the left; we have two arms, two legs, etc. Mutations are random freak accidents, happening one at a time.

11. Does "desire" evolve? Did the male dog still desire the female dog when he was ALMOST a dog? Do body parts and hormones evolve at the same time? What if the hormones evolved FIRST, and the body parts pertaining to them did not exist yet? What if the urge to flex a muscle evolved before the muscle existed?

12. If nature is all in transitional stages, why isn't it all in confusion? Why do we see completed order now? For example, the leaves on trees have patterns of exactly 3, 5, and 7 points; why? Why do we

see every person with two eyes in the center of his face rather than one off where an ear should be?

13. Because apes have jaws that function like humans' jaws, evolutionists use that as more proof that we evolved from apes. Could it not be that we see this phenomenon only because similar structures exist in animals and humans simply because they are functionally the best system to accomplish the task?

14. What selective advantage would there be for deep-sea creatures that don't have eyes to have been made with beautiful colors? What selective advantage would there be for the whale turning into a cow? What would the half-whale, half-cow be good for? How would it survive in the middle stages?

15. Did all of the components on the Periodic Table of Elements evolve? Did the principles of chemistry evolve? How 'bout the laws of physics? Did the power of using a rod over a fulcrum evolve? Yet these absolute scientific concepts are needed for understanding the formation of the animal wing? Huh?

16. How 'bout the difficulty of things evolving frontward and backwards. What if one of a man's feet evolved facing forwards and the other evolved facing backwards? How could he walk? Doesn't this presuppose some sort of genetic code to get them to both evolve FORWARD? If so, where did the genetic code come from? Who wrote it? Did THAT "being" evolve?

17. Now that we know the cell is a complex CITY of biochemical reactions, happening at split seconds, not only is there the question of who or what MADE them but who TIMED them all to work together and not collide.

18. A major question evolution must solve is the evolution of INFORMATION: DNA coding. But, as we see, mutations only happen on PHYSICAL things...not metaphysical things. Does INSTINCT evolve?

19. Does conscience evolve? How do independent consciences, growing by separate and differing rates of "mutations" (of non-physical things, by the way, only evolving ethereal hunches), evolve at the same rate? Some baboons think it is all right to murder their fellow baboons, while other baboons think it is not all right? Some baboons think they should use each other as baseball bats, while other baboons think it is better to eat each other? Do relational arrangements evolve, too?

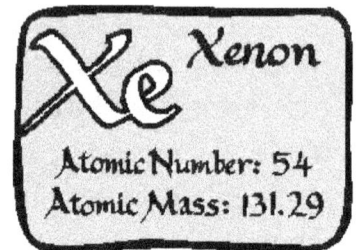

The discoveries by modern science make the theory of evolution downright embarrassing. You might want to review the absurdity of these points with your children and grandchildren.

Could evolutionists be desperately clinging to evolution for moral and religious reasons, rather than scientific reasons?

1. Why have they found it necessary to pad or falsify the fossil record where man and ape are concerned? The Piltdown man was constructed around one tooth and was later found to be a hoax. Often artists add hair (which, of course, would have decomposed by now) to their drawings to further this image of an ape man. The skeleton of an early man who had arthritis could have been bent over and ape-like. Bones in fingers and feet have been arranged to look curled, etc. Why all the effort to create these images? The Neanderthal and Cro-Magnon ape/man hominids are now known to be all man.

2. What do they do with the mathematical probability of chaos producing order? That would be like putting all the pieces of a watch in a shoebox and shaking it vigorously, hoping that when you open the lid a perfectly formed watch will appear.

3. How can time plus chance plus matter equal order and meaning?

4. How could a sense of law and justice evolve? The existence of a justice system indicates that we have developed a system of right and wrong. Where did that come from?

5. How can a brain that evolved from something so low be trusted to decide it evolved? Can an evolutionist trust his judgment now, if he is only half-developed? How could he know to what degree he has developed?

6. How can there be a design without a designer? Could the bottom line be this? An evolutionist thinks that "If God created all this, I'd have to worship Him and that's unthinkable, so there obviously was no designer." He chooses to believe any absurdity so as NOT to believe in any Intelligent Design -- because he might ultimately meet the Designer at some sort of judgment seat.

Is it even possible to be a theistic evolutionist?

How 'bout taking a middle of the road position, believing in a compromise of both? There is no such thing as a theistic evolutionist. A person cannot trust in the Bible and also believe that evolution is the explanation of the variety of fully formed organisms we see. If he attempts to do so, he immediately encounters major problems with his compromised/synergistically held faith.

For example, the Bible clearly says sin entered the world through ONE man, Adam. This does not refer to the entire human race; it refers to ONE man. How do we know? Because this is re-stated in a NEW TESTAMENT passage in the book of Romans...comparing the ONE man, Christ, with the ONE man, Adam (Romans 5:12). Sin CAUSED death. The wages of sin were death. So, death could not have entered the world before sin. Did 100 prehistoric dinosaurs sin? How can we have millions of years of death before even one sin?

What creationists believe:

1. **God created** all plants and animals. Every kind of creature appears abruptly, fully formed. A creationist believes that there are minor changes and adaptations that do appear in plants and animals, but only WITHIN a species.

2. He commanded and ordained that they **reproduce after their kind**--not changing into some other plant or animal.

Circle the number of times "after their kind" appears in this text of Scripture:

> **Then God said, `Let the earth bring forth living creatures according to their kinds, cattle and creeping things and beasts of the earth according to their kinds.' And God made the beasts of the earth according to their kinds and the cattle according to their kinds, and everything that creeps upon the ground according to its kind...**

Could the repetition mean that this was an important point to God? That He never wanted the differences blurred?!

3. God created out of nothing. He created by fiat--using His spoken word alone. God SAID ,"Let there be..."

There are things in creation that are so superbly fine-tuned that they inspire awe and worship. For example, we have already referenced the perfect distance of the Earth from the Sun for comfort and survival.

And another! All liquid forms grow heavier when they turn to solids, except for water. If water grew heavier when it turned to ice, a lake's frozen ice would sink to the bottom and would kill all of the marine life.

4. Creation occurred within a literal six days; He rested on the seventh.
Creation was finished at one point in time. The evolutionist believes, on the contrary, that it keeps going, keeps progressing, keeps reaching new heights, keeps enlarging and creating new matter. This contradicts their own scientific law of thermodynamics. The very first law of thermodynamics is the conservation of all energy and matter. This law states that energy can be transformed in various ways but can neither be created or destroyed. Increasing organization and development is necessary for the theory of evolution to work. Yet we see not one thread of evidence for it.

5. Man was created by a designer for some purpose. The result of a finished creation? Each plant and animal was created with a purpose: to glorify God in a special unique way by being different.

Look up the following scriptures for God's own testimony of His own work of creation:

Hebrews 11:3	Genesis chapters 1-3
Revelation 4:11	Colossians chapter 1
Isaiah 45:18	Psalm 33:6-9
Romans 1:1	

_____

Sources

Morris, Henry M. *The Twilight of Evolution*. 2nd edition. El Cajon, Calif.: Institute for Creation Research, 1998.

Nutting, David. *Fifty Scientific Reasons Why I Think Evolution Is Wrong*. Audiotape. CAPE-NM conference (Albuquerque, N.M.), 1998 April.

Psarris, Spiros. Lecture on astronomy and creation. Durango, Colo., 1999 Dec.

Sorbo, Roger. "A Nuclear Chemist says, 'The More I Study Science, the Stronger My Faith in God,'" *Decision Magazine*, 1999 October, p. 29-30.

A Question of Origins: Examining the Creation/Evolution Controversy. Videotape. Eternal Productions, 1998.

Creation and Evolution: Major Challenges to Darwinian Evolution You Should Know. Torrance, Calif.: Rose Publishing, 1999.

Additional resources

Alpha Omega Institute. P.O. Box 4343, Grand Junction, CO 81502; Phone: 970-523-9943; Website: **http://www.discovercreation.org/**; email aoi@discovercreation.org

Ham, Ken. Answers in Genesis Ministries. P.O. Box 6330, Florence, Kentucky 41022; Phone: 1-800-778-3390; Website: **www.AnswersInGenesis.org/**

Morris, Henry M. *That Their Words May Be Used Against Them*. Master Books, 1998.

Object Lessons

Creation

1. Show an invisible man--either a plastic replica or see-through pages in an encyclopedia.

2A. Circle the times the phrase "**after** ... **kind**" appears in chapter 1 of Genesis.

2B. Examine the seeds in a red pepper. Each has the directions in it to reproduce the plant exactly the same.

3. Show the order and extreme level of organization of a man now, and that he started out fully created.

4. Point out that the Earth is not millions of years old; the shallowness of the lunar dust indicates that it is young; history may be only about 6,000 years old if you trace the genealogy from Adam onward.

Evolution

1. Put all the pieces of the plastic man into a shoebox. Shake it and see if it comes out a man!

2. Demonstrate a stuffed animal lamb giving birth to a stuffed monkey!

3. Cut out a construction paper single cell, and then bang it into some rocks and begin to make dents and holes in it. What is the likelihood that this will become a monkey?

4. Shake the shoebox (from object lesson #1) for a long time. Does it make any difference when you open the lid?

.

Complete the worksheet on the following page. Discuss the moral implications of both views for points 5, 6, and 7 on that page.

Worksheet

Creation

1. There was a _____.

2. In the _____, it says

   "Everything is to reproduce

   _____ _____ _____.

   How many times? _____.

3. We started _____ _____--

   _____.

4. It took _____ days.

5. Since you were _____ by _____,

   you should _____ _____

   neighbor.

6. There's _____ quality control

   Life is _____. Let

   everyone _____. Only

   God is _____ to

   _____ life.

7. We have a _____. We

   were made to _____ God

   and to _____ Him _____.

Evolution

1. There was _____ designer.

2. Plants and animals _____.

   They come out _____

   with each _____.

3. We started from __

   _____ _____.

4. It took _____ years.

5. Since you're an _____,

   it doesn't _____ how

   you _____ your neighbor.

6. There is quality _____ .

   Throw out the _____

   babies and the _____

   old _____.

7. We have _____ purpose.

   Just go do _____.

   _____ your sister.

   Make _____.

   Seek _____.

Worksheet *Answers*

Creation

1. There was a __designer__.

2.. In the __Bible__, it says

   "Everything is to reproduce

   _after__ _its_ _kind__.

   How many times? _ 10_.

3. We started __fully_ __created__--

   __eyes in the right place, etc.__.

4. It took __six_ days.

5. Since you were __made_ by __God_,

   you should _love_ __your_

   neighbor.

6. There's _no_ quality control

   Life is __sacred__. Let

   everyone _live___. Only

   God is _allowed__ to

   take life.

7. We have a __purpose__. We

   were made to __glorify_ God

   and to __enjoy_ Him _forever_.

Evolution

1. There was __no___ designer.

2. Plants and animals __change__.

   They come out __different_

   with each _birth____.

3. We started from _a_

   __single___ __cell__.

4. It took _1,000,000_ years.

5. Since you're an __accident_____,

   it doesn't __matter_ how

   you _treat_ your neighbor.

6. There is quality __control__ .

   Throw out the _bad__

   babies and the __weak_

   old __people____.

7. We have __no_ purpose.

   Just go do __drugs____.

   __Hit__ your sister.

   Make _money_.

   Seek __pleasure__.

ABOUT THE AUTHOR

Renee Ellison's practical tips have been a boost to thousands of moms. With over 30 years of experience in education, Renee brings a rare understanding and an easily comprehended synthesis to each of her publications. She has taught nearly all grade levels and was an elementary principal, head of a high school English department and Teacher of the Year. She developed and ran gifted and talented programs and has conducted numerous teacher training workshops and seminars at conferences in North America. You'll finish reading her books with new-found understandings. Renee's unique knack and wit for making complex things simple makes reading her materials and using her phonics program and her typing and piano keyboarding courses a pure delight.

+ For blog posts, podcasts, and much more, visit www.homeschoolhowtos.com

+ Visit Renee Ellison's YouTube channel

www.ingramcontent.com/pod-product-compliance
Lightning Source LLC
Chambersburg PA
CBHW051422200326
41520CB00023B/7327